the american poetry series

Vol. 1   Sandra McPherson, *Radiation*

Vol. 2   David Young, *Boxcars*

Vol. 3   Al Lee, *Time*

Vol. 4   James Reiss, *The Breathers*

Vol. 5   Louise Glück, *The House on Marshland*

Vol. 6   David McElroy, *Making It Simple*

Vol. 7   Dennis Schmitz, *Goodwill, Inc.*

Vol. 8   John Ashbery, *The Double Dream of Spring*

Vol. 9   Lawrence Raab, *The Collector of Cold Weather*

Vol. 10   Stanley Plumly, *Out-of-the-Body Travel*

Vol. 11   Louise Bogan, *The Blue Estuaries*

Vol. 12   John Ashbery, *Rivers and Mountains*

Vol. 13   Laura Jensen, *Bad Boats*

Vol. 14   John Ashbery, *Some Trees*

Vol. 15   Sandra McPherson, *The Year of Our Birth*

Vol. 16   John Crowe Ransom, *Selected Poems*

Vol. 17   Robert Hass, *Praise*

Vol. 18   James Tate, *Riven Doggeries*

Vol. 19   Dennis Schmitz, *String*

string

the ecco press / new york

string / dennis schmitz

Copyright © 1976, 1977, 1978, 1979, 1980 by Dennis Schmitz

All rights reserved

First published by The Ecco Press in 1980

1 West 30th Street, New York, N.Y. 10001

Published simultaneously in Canada by

Penguin Books Canada Limited

Printed in the United States of America

The Ecco Press logo by Ahmed Yacoubi

Typography by Cynthia Krupat

Library of Congress Cataloging in Publication Data

Schmitz, Dennis, 1937–   /   String.

(American poetry series; 19)   /   I.   Title.

PS3569.C517S8   811'.5'4   79-9444

ISBN 0-912-94668-7   /   ISBN 0-912-94669-5 pbk.

Grateful acknowledgment is made to the following publications
in which these poems first appeared: *Antaeus*: "A Rabbit's
Death," "Arbeit Macht Frei," "Dog," "Making Chicago," "Navel,"
"Planting Trout in the Chicago River," "Soup," "String," "The
Other Side," "Trying to Remember Paradise," "Wings."
Back Door: "Conjugal." *Field*: "City Ginkgos," "Delta Farm,"
"Infinitives," "Mile Hill," "The Feast of Tabernacles," "The Man
Who Buys Hides." *Iowa Review*: "Divining," "Homing," "In
Bosch's World," "Making a Door," "Rabbits." *Poetry Now*:
"Hard Labor," "The Poet at Seven."

I would like to thank the National Endowment for the Arts and
the John Simon Guggenheim Foundation for grants which made
completion of this book possible.—D.S.

for anne, sara, martha, paul, matthew

contents

one
___

delta farm : 3

divining : 8

infinitives : 9

homing : 10

making a door : 11

rabbits : 13

mile hill : 14

conjugal : 15

a rabbit's death : 16

the man who buys hides : 17

two
___

string : 25

navel : 26

dog : 27

soup : 28

the other side : 30

trying to remember paradise : 31

leaving the palace : 32

the poet at seven : 34

hard labor : 35

arbeit macht frei : *36*

in bosch's world : *38*

city ginkgos : *40*

wings : *41*

the feast of tabernacles : *42*

planting trout in the chicago river : *44*

making chicago : *46*

one

delta farm

a friend weighs little a wife
makes the body heavy
as she swims away in the marriage
sheets—she seems more
strange than my mother's

face surfacing
in memory. so the drowned
displace the living—
not my wife's but
mother's thirst dries the sweaty

fingerprints
from the handle of the short hoe
or cutter
bar skimming the overflow
our salty bodies deposit between
windrows. together we pressed

drool from the sugar
beets & threw
or wished to throw our bodies
like pulp to the few
hogs we kept for meat on the tufted
mud of an upstream
island. this is the sweetness we refused
one another.

.

this neighbor is married
to solitude
another whose bridal sheets still smell
fresh drinks
greenish mouthsful from the cistern
children won't grow
up their roots churn in
the cultivated
zones fractured by a hundred
canal reflections.

when women visit
they only fix cots in fallen
down coolie shacks below
the town produce
sheds now abandoned & shifting
with the sun's weight.
when they leave boys

will lie restlessly
fishing in the narrow
beds skiffs make, between the pilings
hear the sheet
metal pop nails to trail
in a swifter river.

late November: a sixty-knot
squall through

the Straits breaks
levees, backs salt water miles
inland to preserve
what it kills. animal features
wake on the bedroom windows—
buck deer the flood divorced;
our cow sewn with scars
bawls, against the dawn rubs
her painful ballast

of milk. my wife by instinct
washes her own
breasts before our daughter
feeds. birds in refugee dozens
scatter as I walk
a smaller yard. for days only boats

define the horizon. only the doctor
salt-stained
like us in boots & overalls
scares us. our daughter crawls
through fever one week
then her mother the week after

dies. my wife,
still my wife, what I have
of you, this residue, this love-

salt, will not let me cross private
places in my body
anymore. without you
I can only continue a snail inside
my shell of sleep trailing sticky
dreams. nowhere to walk I go
away from you.

.

I am forty-two, my body
twenty-five.
my skin recopies over
& over my small daughter's

hand barely
holding against its current.
once I wanted to be still
water, a puddle the sky
fell in or halo
my forehead molds from the saturated
marsh when I bend

my face to the first
unconfirmed rice shoot.

now I wait for the March-fed
river to clean
the delta, knead our thought-
out acres of orchard high
ground where picking ladders descend

legless into their own
reflections. the bottomland rice
is lost but these trees
reach deeper. rings of salt show
each step back the sea
takes. swaying from tree
to tree at last my daughter learns

to walk.

divining

      small world, the intricate
root protruding: imagination that makes
orchard trees
      so close they knit

inevitable distortion.
what then is the sky but delayed shadow
the eye a failing

      source of light?
after memory what leaf
won't seem faulty—witness poor van Gogh
at Arles sketching real
orchards until "ideas were the eyes
      of the eyes" & the dark of the mouth,

damaged night. overhead on my work-
shed the gross acacia
      crawls the metal
roof though I thin
it—the West is angular & cerulean
in the limbs where the bow saw
      still hangs teeth
up.

infinitives

first speech—no begots, but
born of itself, starts:
the baby crawling the yard also
crawls "bird," "stick,"

or "tree." to be is to be raised
from mother flesh plural,
to love distinction but be some part

of travail that summers
in an older body. Sara's grandmother
speaks as she walks
the sunburned orchard holding Sara:
these are the limits
& so is the way you repeat

them. you
drink this truculence; if you don't
choke you are healed.
sometimes it is only for a season
the eye must be a halo
for discolored April & you hold

back speech the way small carp
clog the irrigation
pump as it pants raising slough

water.

homing

           6 pm sun
condenses on the unequal sea.
I imagine Tomales sharks
twitching the chill
with rough flanks, homing,
      & the flounder their prey damned

to stare upward, homing.
you taught me to accept my scent,
how to speak face down
the slow zeros of exhalation—

in short, to swim against what
seemed to me human.
I was wrong, not wronged
not to try that other water,
proceed to the difficult

backstroke over a cobble of pool lights,
my eyes upward, the insatiate
lungs pumping into darkness.

making a door

a weedy creek
peeled from cornfields,
the whole countryside
where I grew up
thaws from the front
windows of this dollhouse

we are making together.
my daughter kneels
to chalk night
on the back windows,
wanting for this one house
all that our family lived
her eight years—
even dreams reduced
to the neat minimum

of her bedroom.
I ask to enter
the doll's world,
tell in altered size
what I dreamed

in my half of the house:
how I reached speech
through a series of dahs,
made my face a welt
on the five senses—

I go on distributing
myself over the assigned parts.
the house is almost done.

I hand her the saw.

rabbits

the urge to stroke the dead
    one back, hand-feed life
to the animal
body even as its soft
    vision dilates,
your calluses pulling
fur like lint from the unmendable

flesh. shake your head
    & coming back, hose the hutch
before your wife
six months pregnant sees

the rabbit. later
    she can launder your sweaty
overalls & empty the few
    black rabbit
pellets your pockets
caught. in the closet you

change & relishing the bachelor
scents of your underwear
    drop it to your father-in-law's
bathroom floor. now your voice

weighs nothing though
    you sing.

mïle hill

December: the trees chafing.
instead of a hole
at the horizon the focused light
of a welder's torch: the sun

& the iridescent this-world fuse.
6 days' drive, Calif to the cramped Iowa
farms. by the roadside we stretch
as I explain where my family

grew. below,
small preserved Dubuque bristles
in '90s plain-face
brick across the uneven hills,
circles where the river does

south to slough water.
Sara picks up
snow; molds it to her small hand,
tinges it with her pink

flesh: concomitant beauty
the bloodspot on the egg
we are on our knees
everyday to find on the ground
what we'd lost to the sky.

conjugal

after morning rain in blue cobweb the sun
kindles. in every window
over the yards steeped faces bob

up: children first, instinctual,
reforming as they breathe
in sour zeros, round mouths
sucking the other side

of glass. the nose, useless
dorsal, then other muscles knit
the face around
obsessive but artless patterns.
outside they see us

framed & purposeful
because we love in others what is
most foolish in ourselves,
sense or scenario—as the outsiders

might have seen
when the rain first blistered
the windows across
our own domestic shifts,
my face to yours
the focus of surprised longing.

a rabbit's death

the rabbit's spine must be snapped
by the left (thumb down)
hand as the right cradles the furious

body. the left hand, left
foot gauche because
nonliteral & out of symmetry

suffers with the disconsolate
humor of your species.
     what can the hand repent,
the eye
& mind abase? this rabbit's

death, this contagion blessed
by being altered to man
     who must die also
kicking? so fury apes gusto,

its gratuitous opposite.
     after the back legs are broken
at hock
one finger underneath can lift the skin
whole to be pulled
off entire with the twisted

head attached.

the man who buys hides

before I had a face
my mother supposed another horizon
aligned with that

wrong one contemporary hands fit
to relic Sioux.
maybe town behind the long
hills her brother's feedlot rode.
she was unmarried & wished

my face would never ripen
on the small stone memory left
of a necessary
lover. but I was born

white. & grew, bulky
& slow, never hearing my name
knotted in her tongue though
she must have sung

kneading father's pale
flesh, face to face drinking
distortion from each ripple
the other's body made.
or so I thought
when at night her lost
voice sounded the walls to the service

porch where at fourteen I dreamed
one pool dropping

to another hatchery trout threaded.
on the wall overhead
an exhausted
Jesus dried in milky shellac,

showing rain.

        .

summers I answered
her face surfacing in the bedroom
window to clean her eyes

or devotedly follow a crippled
hand through the word
"drink." my denim printed
blue sweat wherever I leaned
in her unquenchable

shadow. winters the ragged sky
my eyes folded as I slept
lost snow. the world formed under

our stamping feet & above it
breath drifted. with bare

hands uncle & I
shaped the cattle's frozen
noses, undid the ice their drool
tied through whatever
they ate. only the radio

spoke other names
when the lights were out
& eyes were adrift in their own

local winter.

.

forty years later never thinking
of Dakota I still go faceless
into sleep & dream myself

intaglio under the animal.
I never see the driver
of my truck as it weaves the dead
smells through eucalyptus rows
swerving for potholes

in the gravel till the bones
give & under the tarp
the burst organs suck & squeeze.
what is death?
at the tail end they put one drain

at the other end they put
the tongue back in, intractable
& too big—
what can the doctor say

who is tired of his own body?

                .

the poplar leaves go on multiplying
basic July. July sun
is swollen in the basin

where I cool
my drunken face. once I loved
my smell as I loved the hoof tipped
with stink trailing from the stud

barn. I have become a talker
in bars who wanted
to be only a handspan of red
earth trembling with ants.
Dakota stays under the washed
face even if Calif
turns it dark. only man is dumb

whose tongue shapes before the fingers
know. did these dead
animals talk

with hooves or with tactile
fetlocks praise
the grass as they stepped

off the limits of their hunger,
touching with their mouths
last? as I load each distended

body the winch
squeals, the cable cuts new
boundaries across
a piebald hide. only the head
drags & the eyes roll

over, counter to the earth.

two

string

no one knows the way out of his mother
except as she leads him
the final knot is his head
which drags the nerves
along the spine like doublestitch
up the reverse
of the body & outside the soft fabric
currents which pucker
erogenous zones. no other joy

but this string, this dorsal string
one end shit, the other end tongue.
who has not asked the way back?
who is not guilty of graceless longing,
& alone? watch that man who

puppet-dances through noon
traffic skinning light
off the chrome. though his hands
are broken his quarrel
goes on. his pants front soaked, shamed,

he sings for pedestrians.
his mouth will bunch with old stitches
but he will sing, "privacy is only

contraction, heavy
body, dangle of shriveled nuts . . ."

navel

older than the anus with which it shares
alternate privilege
north pole to south pole—
even the face is a navel
transplant which never

takes root. or pretend it is an eye
blotted at belt level
point it at your wife & she
at you before you part
snoring, your belly button receding

to a blind knot. but we who tie
kisses in our own flesh,
hypnotize ourselves with memory
until hate stares back,
if we finally marry will the tooth
break when we bite those many
faces the muscles form
over her belly as she shudders

with love? what other way
can the mouth tattoo
meaning on wary desire
the tongue swim its sour
journey into the thighs running with sleep?

dog

after they had sewn him into the dogskin
they called him dog
as they had called him bird

    when the sky contracted.
to be human is to ache openmouthed—
the voice flapping, helpless.
can one say, without smiling:
    this is Gregor the peripheral

roach, a million years new,
& still hug man, dignify his pitiable daily needs
by existing in equal pain?
you would make yourself a dog

    first & pee openly, delighting
in this pliable flesh so cursed
with particular life that what seems gentle
    sweetens the repetition.

soup

a fistful
of spinach or chard sleep will make
a garden, & hidden snails that dying whistle

as the bunch opens, boiling: cautery.
temporary drunks, or just jobless
you sing as envy cools—
one voice for bread, full of yourselves,
queued for bowls & the spotted

flatware I wipe & count
as I give them out.
across Halsted a second
kitchen feeds the crippled, the old,
those who never dry out

who hum in common
speech Chicago, moan Chicago.
I'm the one I haven't counted
who won't eat from your bowl
though I wish humiliation for myself

rather than anger.
each time, I go away undefined
though once I threw up—

breathing against my own fouled mouth
I asked you to forgive me

to stand aside as I knelt
sick with what I could not consume.

the other side

going blind the eyes at first
are stained with
        the world below & the stunned tongue
wants to grip
anything for memory—dog-ear
        leaves high
in the dipping elm your telephone wire
cuts through. you can't forget

your sister's voice,
        as you ascend the wire, saying
you are a child afraid to cross sleep.
the other side
        is immense August. you balance
carefully to drop
        clothes. on one leg, you kick
off the opposite

        shoe. body fat swaying
you practice before family,
        eyes tied over with a stocking,
until being naked no longer bothers you
        & you let go
because the hands are exact
        over things
you love for a long time & the feet
go their own way.

trying to remember paradise

the placenta all dead fizz & precious
mask, matrix your face
pounded to a twin your mother

might have loved better
is the one you blame
when your body is too heavy for sleep.
you rock in the counterpoint

of another's fatty heart;
you are the thin lip of muscle a clam
uses to close its elaborate identity.

who was it you killed
for Abel's gifts—an animal Adam
never named, a martyr
gilded by his own greasy fire?

when God pulls the delinquent
ghost out of your
lungs, whose nasal song is it?

leaving the palace

in the arcade cut from
the '40s theater too big for movies
behind butcher's glass
undulate & deep enough for faces
you can sense snouts

inside bins, other parts looped
over hooks—a wrist dwindles
to a knot, kinks take up

the maculate purity of flesh.
a pet nook folded from sheet metal
the shape of a heart, pretzels
on a stick sold by a one-armed man who

will tie them with his hand
concurring in the pattern you request,
spell your girl's name, or the spoken

approximate of love. you bargain,
for a few dollars you can even buy
the separate organs for good & evil
have them sewn
under your clothes & afterward walk

lopsided, mongrel saint,
by rote damning men but praising tedium
that betrays them

to stalls where the voice is spilled
into the grunt & whir
             of machines.

the poet at seven

for John Logan

the boy loses his way
because he wants to, because the dreamer
always has the role
of DESPAIR while the other children
mistakenly wear their father's clothes inside
out, or the girls work intricate
knots with mother's underclothing.
the boy's voice deepens
in useless dialogue with plants
he can imitate. what is the word
the tongue evolves to? already the hands
are solo when he draws
the two-legged sheep in his account
of heaven. already he can lift
his head from its print
in dreams & soon he will learn
the eye, even when released
into sleep, is just a fish the steady
river forms.

hard labor

my knees form around the emphasized
belly, the parenthetical
child flexing in the first
human pleasure. for my husband I was

a seat made of lover's poses.
but his delayed face
won't refocus on the smile lines
in my thighs—I raise a leg
& the muscles creep
into remembered distinctions.
one with a rat's nose pokes behind

my back until I groan monotonous
as dial tone. no one hears,
no one celibate knows
this singleness I can't undo
though every nerve-knot repeats

the formal twist & pull
of our love-making.
even between my legs this woman's
halo which will tighten
around the baby's head begins

to clench its own way.

*arbeit macht frei**

his first day they asked
him to remove the swallows
so he hosed the mud
nests high up under

the sign while we watched, hats
off. once such ladders ran
out above a man's house
to touch frayed clouds.
now, a man is hired to wash the sky

or urge grass.
if you are uncomfortable in our world,
stop for a minute & put your eye
to the entrance hole, the swallow's
exit, hose pointed to another

nest. you look
inside: a woman is sprawled, legs
wide. you peer into her

hole too & see
a third hole or original
of your mouth hatching

* *"Work sets you free"—inscription on the gate at Auschwitz.*

(36)

the word "human,"
pretty feathers the spit
fades. press your face to the nest
& reach out with your liberated
tongue remembering
other words it kneaded into love
as many climb the ladder
behind you.
underneath, mud darkens

the tidy yard.
is that a guard kneeling
by a hole accidently washed
from a nest? as he reaches down
the hole closes
over his ring finger.
this is how birds are made.

in bosch's world

too we are an odd detail—
a man & wife riding a fish
into the sky, too incidental to be

St. Anthony's temptation to carnal knowledge.
we refuse to look at each other
or at the scenes that encourage us

to suicide, ignore the street boys
beating a dead man's face,
the sum of carnal knowledge.

here men can have the faces of turtles
with the accuracy of hard practice.
all work on the body piecemeal:

one man develops the hand,
or accomplishes only *adductor pollicis*.
each man is still only a piece of man;

this is the village, the first Chicago.
in such gestures as allegory permits,
guile is one-handed, but self-disgust

has cut off both its hands.
what made cities, civility, not even
kindness, is broken to fingerbones,

to nail parings a later man will realign
with tweezers, & with regret at the waste,
make only a handshake from the bones of two bodies.

we are leaving whatever way we can.
below, Anthony can go on kneeling
to his own demons, unable to cajole Jesus,

to make him fly. he may want to be elsewhere,
he may want to ride the pony who
rubbed his saddle off on the bare wood

behind the picture; not to be hero
or saint, to dream humble dreams
lit by the spit-sequins glazing

the pony's muzzle as he labors to be.
in this Hell vermilion masks the suicide's
face & habitual ocher the hero's backside—

the kind of hell we too could have made up
separately.

city ginkgos

                    our kin,
cyanotic, their voluble
crowns heavenward, root in the immediate

concrete. if I could speak ginkgo,
its verbs excepted,
I'd show our Chicano neighbor's blue
face, some sky pigment, as he hangs himself

for the nights his wife slept elsewhere.
the ginkgo nouns suffice but its flimsy
limb lets go & he comes down

in his own yard, guttural but alive.
things safe to say in English we say
when we meet—
no ironies, no euphemism briefly
dipped in pain, but things
dissolved in the senses that can't

come back words.
I can pretend Spanish, I can pretend
enough to moan
into intelligence his wife's lament
living with a man, you talk & talk
until the tongue
detached from words crawls

out of love.

wings

I am the carpenter called
three times in a dream drooled like resin
out of our marriage sleep,
who peeled the raw sun that bewilders

owls, who began with the birds
I pressed & dried as a child
in heavy nature books, who now ties
the angel's featherless arms

to this dreadful trellis.
yesterday's sky is also trained
to zones between the laths
as on each wing I work the truth

& doubt incised on our own hand-
prints. I want truth to repeatedly feather,
as in my dream, from the right—
but when I let the angel go

I want him to fly in ever tighter circles
with the stronger, more human, doubt
fanning our upturned faces.

the feast of tabernacles

> . . . all that are Israelites born shall
> dwell in booths: that your generations
> may know that I made the children of
> Israel to dwell in booths, when I brought
> them out of the land of Egypt.
> —LEV. XXIII, 42–43

no Chicago like this
the blacked-out Scouts give themselves
as they cinch tents to the three

elms inside the fenced lot we call
Elastic Acre, cramped by 3-flr walkups:
sons of the sons of Polish Jews
born to Chaldeans, or to an Egypt

that squeezed them
out between the stitches broken
in bulgur sacks, manna
of La Fortuna grits & mealy worms.
the forest they make up
twig by twig is less wild than the city
they will to forget.
their last knots are done

by flashlite, the fire lit
out of an ignorance learned from books.
if their songs blur
as I sing them so that the forty years

their fathers wandered fit
my forty years thrown
Moses-size across their tents though

the light inside
is small, their songs still surprise God
in them, still loosen the mouth
from biting . . . but better this flesh,
this bite which outlines flesh,
than the food of heaven,
the peeled herbs, locusts, or any
popcorn allegory of the wilderness

planting trout in the chicago river

fishers of men
—MARK I, I7

because it is lunchtime
we lean from offices high over the river.
at a holding tank below the mayor
is on his knees as he mouths
the first fish coming down the hose.
his wife pulls back crepe
so aldermen & ethnic reps can help themselves

to rods as he bends
over the awful garbage, as his lips let go
this fish rehearsed four nights
in which his tongue grew

scales, the Gospel epigram hooking
his mouth both ways
until he talked fish-talk
but still hungered for human excrement.
even in sleep he can will the river:
the midnight bridges folded

back for a newsprint freighter,
the Wrigley Building
wrinkling off the spotlit
water—he submerges a whole city

grid by grid to make one fish.
four nights he dreamt the different
organs for being fish,
but out of the suckhole soft
as armpit flesh a face always bobbed,
on the eyes a froth of stars.
in turn each of us

will have to learn to cough
up fish, queue from Lake St. Transfer
at the river to disgorge
& after, wade at eye level

under Dearborn bridge, intuitive,
wanting flesh, wanting
the warm squeeze & lapse
over our skeletons because the fin repeats
fingerholds, the gills
have an almost human grin.

making chicago

We cannot take a single step toward
heaven. It is not in our power to travel
in a vertical direction. If however we
look heavenward for a long time, God
comes and takes us up. He raises us
easily.

—SIMONE WEIL

let it end here where the blueprint
shows a doorway,
where it shows all of Chicago
reduced to a hundred prestressed floors,
fifty miles of conduit & ductwork

the nerve-impulse climbs to know God.
every floor we go up is one more down
for the flashy suicide, *for blessèd man who*
by thought might lift himself

to angel. how slowly we become only men!
I lift the torch away, push up the opaque
welder's lens to listen for the thud
& grind as they pour aggregate,

extrapolating the scarred forms resisting
all that weight & think
the years I gave away to reflexive anger,

to bad jobs, do not count
for the steps the suicide
divides & subdivides in order not to reach

the roof's edge. I count the family years
I didn't grow older with the stunted
locust trees in Columbus Park,
the ragweed an indifferent ground crew
couldn't kill, no matter the poison.
now I want to take up death more often

& taste it a little—
by this change I know I am not what I was:
the voice is the voice of Jacob
but the hands are the hands of Esau—
god & antigod mold what I say,

make me sweat inside the welding gloves,
make what I thought true turn heavy.
but there must be names
in its many names the concrete can't take.
what future race in the ruins
will trace out our shape from the bent

template of the soul,
find its orbit in the clouded atmosphere
of the alloy walls we used as a likeness

for the sky? the workmen stagger
under the weight of the window's nuptial
sheet—in its white reflected clouds
the sun leaves a virgin spot

of joy.

*Dennis Schmitz was born in Dubuque, Iowa,
and educated at Loras College and the
University of Chicago. He teaches at
California State University, Sacramento.
Schmitz has received grants from the National
Endowment for the Arts and the Guggenheim
Foundation.*